© 1990 Franklin Watts

Franklin Watts Inc
387 Park Avenue South
New York, NY 10016

Printed in Belgium

Designed by
K and Co

Photographs by
NASA
Johnson Space Center
TASS
Novosti Press Agency
JPL

Technical Consultants
L.J. Carter
S. Young

Library of Congress Cataloging-in-Publication Data

Barrett, Norman S.
    The picture world of astronauts / Norman Barrett.
        p.   cm. — (Picture World)
    Summary: Describes the training, preparation, and kind of missions
faced by astronauts and discusses the pioneers of space travel.
    ISBN 0-531-14053-9
    1. Astronautics—Juvenile literature. 2. Astronauts—Juvenile
literature. [1. Astronautics. 2. Astronauts.] I. Title.
II. Series.
TL793.S735 1990
629.45—dc20   89-21523
                                                               CIP AC

The Picture World of
Astronauts

N. S. Barrett

CONTENTS

Franklin Watts

New York • London • Sydney • Toronto

Introduction

Astronauts are men and women who travel in space. In the Soviet Union they are called cosmonauts.

Living and working in space calls for special abilities — a quick, clear-thinking mind as well as physical skills.

Space travel is dangerous and expensive. Flights are planned years ahead. Astronauts go through intensive training before they fly in a spacecraft. Each mission is thoroughly rehearsed.

▽ The crew of a space shuttle pose for a preflight picture. These astronauts, all highly skilled at operating the complicated systems of a spacecraft, flew on the seventh shuttle mission. As with most space flights, they were making history — the first five-person launch.

△ Wearing a special spacesuit and backpack for maneuvering in space, an astronaut performs tasks outside the spacecraft.

▷ Cosmonauts aboard the *Mir* space station perform a "concert" as they float in their living quarters. People and things in space are "weightless" because they are no longer affected by the Earth's gravity.

The pioneers

All space travelers are pioneers. They are the explorers of the modern age. They risk their lives as they test new spacecraft. They perform new tasks in unfamiliar situations. They set foot on new worlds.

Ever since cosmonaut Yuri Gagarin became the first person in space in 1961, brave men and women have carved their names in history. Some, tragically, have lost their lives.

▽ Soviet cosmonaut Major Yuri Gagarin about to embark on the great adventure. Strapped into his spacecraft *Vostok 1*, he was rocketed into Earth orbit on April 12, 1961, the first human being to leave the pull of the Earth's gravity and enter space.

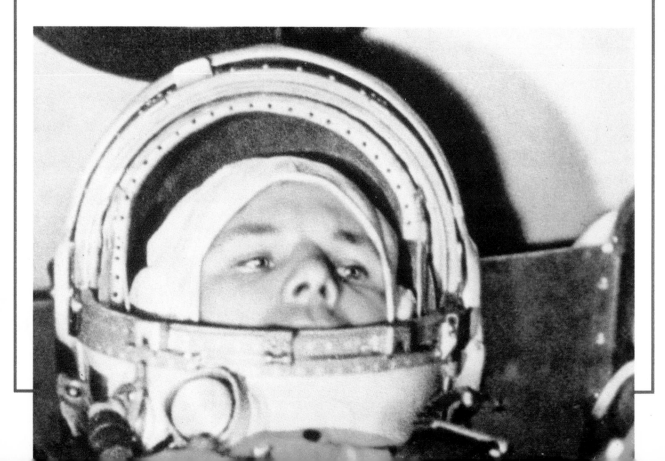

△ Alexei Leonov, the first man to walk in space. He left his spacecraft *Voshkod 2* on March 18, 1965, and spent ten minutes outside, tethered to the craft.

◁ John Glenn, the first American in orbit. The first manned American flights were in Mercury capsules. Alan Shepard and Gus Grissom both made brief suborbital flights in 1961, before Glenn stayed up for nearly 5 hours, making three orbits, on February 20, 1962.

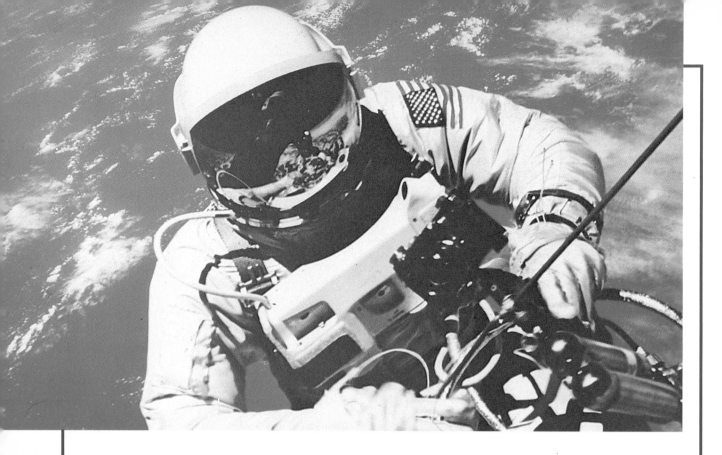

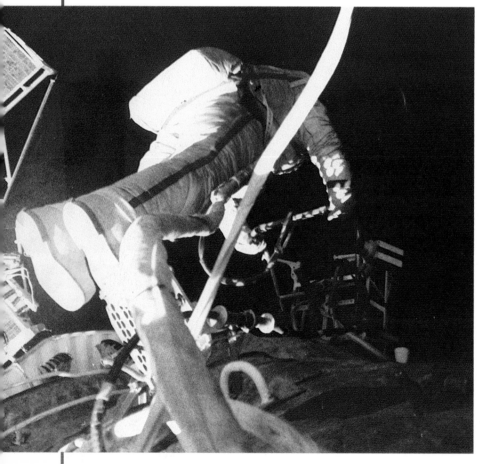

△ Astronaut Ed White, with the Earth in the background, making the first American spacewalk. He stayed outside his Gemini spacecraft for 20 minutes on June 3, 1965. White died in 1967 with two other astronauts when *Apollo 1* caught fire on the launch pad.

◁ Cosmonaut Svetlana Savitskaya outside the Salyut space station in July 1984, when she was the first woman to walk in space.

▷ Neil Armstrong, the first astronaut to set foot on the Moon, or any world outside our own, on July 20, 1969.

▽ John Young (left) and Robert Crippen, with a model of the space shuttle, made the first shuttle flight in April 1981.

▷ U.S. astronaut Bruce McCandless becomes the first human satellite, orbiting the Earth by himself. This was the first untethered spacewalk, made on February 7, 1984.

First ladies of space: ◁ Valentina Tereshkova became the first woman in space on June 16, 1963, in *Vostok 6*. ▽ Sally Ride became the first American woman in space as a mission specialist on the *Challenger* space shuttle in June 1983.

▷ Astronauts are lowered into a pool for special training.

▽ Underwater, with the help of divers, they practice maneuvers they will use in space.

All in a day's work

When a spacecraft such as the shuttle goes into orbit, the astronauts have a whole program of tasks to complete.

The commander and pilot might use the shuttle's engines to change orbit, or the thrusters to make minor changes of position.

Mission specialists usually have satellites to launch, and work inside and outside the craft. Payload specialists carry out experiments.

▽ An astronaut, attached to the manipulator, or robot arm, launches a huge satellite back into orbit after completing repairs. Satellites are normally released from inside the shuttle automatically or by use of the robot arm. Working outside a spacecraft is called extra vehicular activity (EVA).

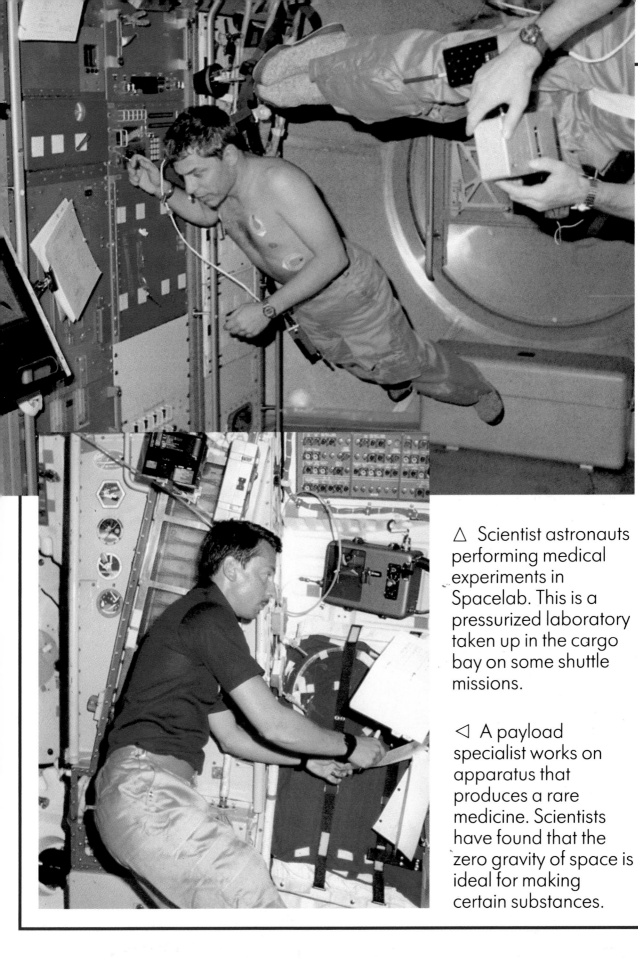

△ Scientist astronauts performing medical experiments in Spacelab. This is a pressurized laboratory taken up in the cargo bay on some shuttle missions.

◁ A payload specialist works on apparatus that produces a rare medicine. Scientists have found that the zero gravity of space is ideal for making certain substances.

All astronauts are given basic training in EVA. Spacewalks are usually performed by the mission specialists.

Most planned EVA tasks on the shuttles are connected with satellites. Construction work on space stations will also call for considerable outside activity.

Unscheduled EVAs might include emergency work on satellites or on the cargo bay doors.

△ Astronauts in the cargo bay of the *Challenger* shuttle work on a failed satellite after retrieving it from space.

▷ Attached to the robot arm of the *Atlantis* shuttle, an astronaut checks joints on a tower that was built and dismantled as part of tests on construction work in space.

Spacesuits

Spacesuits are burdensome to wear and require special training before use. They are worn over a one-piece cooling garment.

The legs, including boots, are put on first. The rigid body section, made of aluminum, clips on to the waist ring of the lower section. It has a built-in life-support backpack. Gloves and helmet are also sealed by snap-rings so that the spacesuit is airtight.

▷ An astronaut, fully suited up, works in the shuttle cargo bay. Under his helmet is a "Snoopy hat," which contains headphones and a microphone for two-way communication. His backpack contains enough oxygen for six hours' EVA.

▽ Getting in and out of a spacesuit is something astronauts practice in training.

For working in space unattached to a spacecraft, astronauts use a special Manned Maneuvering Unit (MMU). This snaps on to the back of their life-support system.

The MMU enables the astronaut to move around in space by firing gas jets. It looks a bit like the back and arms of an armchair.

The astronaut flies this personal spacecraft by operating handgrips on the armrests.

▽ Using his jet backpack, an astronaut maneuvers in space close to the shuttle.

Living in space

Living in space takes some getting used to. The simplest everyday actions on Earth, such as eating or going to the bathroom, become special problems in zero gravity.

Most of these everyday living problems have been solved. But living in space for long periods of time presents a different kind of problem — the effect this has on the minds and bodies of the astronauts.

△ There are no floors or ceilings in space. This is what working in a space station will be like. Scientists carrying out experiments in zero gravity will themselves float freely inside their laboratory.

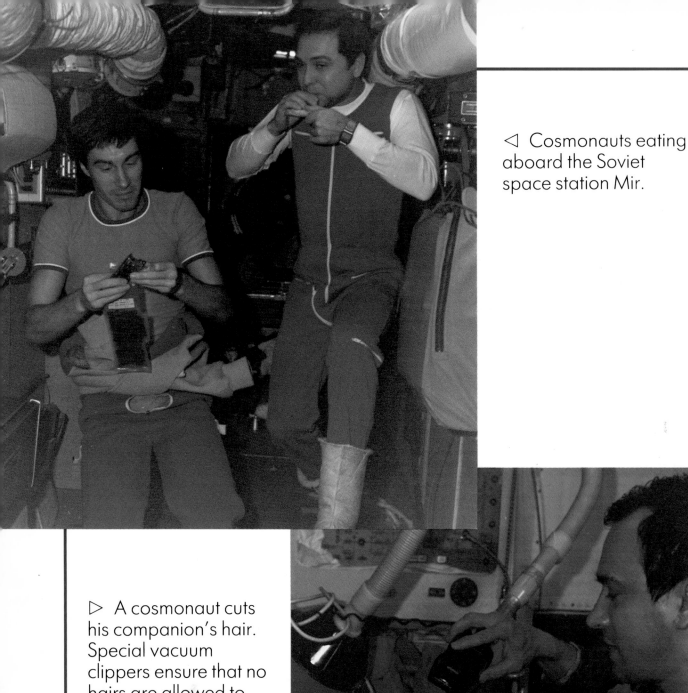

◁ Cosmonauts eating aboard the Soviet space station Mir.

▷ A cosmonaut cuts his companion's hair. Special vacuum clippers ensure that no hairs are allowed to float around in the cabin.

The Russians have been testing the effects of long space flights on the human mind and body. Some cosmonauts have stayed up for as long as a year.

The effects that the low gravity of space has on the human body include the wasting of muscles and the loss of minerals in bones, which become thinner. Much more research is needed before astronauts can go, say, to Mars, a trip that would take years.

Just as important is the effect that such long periods of isolation would have on the mind. Special tests on Earth as well as in space will show how the human mind reacts.

▽ A glimpse of the future – a human colony on Mars. Before the first humans even land on Mars, astronauts will have to make longer space flights than any yet experienced. They will be the new pioneers, taking further steps into the unknown.

Space records

Soviet cosmonauts Vladimir Titov and Musa Manarov set up a space endurance record of 366 days. They were launched on December 20, 1987, arriving at the Mir space station two days later. They returned to Earth on December 21, 1988.

Another cosmonaut, Yuri Romanenko, who had held the record with 326 days, logged a record 430 days in 3 missions.

△ Soviet flight control monitors the meeting of cosmonauts after the docking of a Soyuz spacecraft with the Mir space station. The French flag is in honor of one of the Soyuz cosmonauts, Jean-Loup Chretien, who became the first nonSoviet nonAmerican spacewalker. He returned to Earth in December 1988 together with cosmonauts Titov and Manarov after their record 366-day stay in space.

Most space trips

Astronaut John Young set a record of six separate trips in space. He flew twice in Gemini spacecraft, twice in Apollo, including a Moon landing, and twice on a space shuttle.

Young commanded *Apollo 16* and spent 20 hours exploring the surface of the Moon. He commanded *Columbia* on the first and ninth shuttle missions.

Volunteers for Mars

Compared with Soviet research, NASA has not carried out such extensive tests on astronauts in space. The longest American flight is the 84 days spent by Gerald Carr, Edward Gibson and William Pogue aboard *Skylab*, an experimental space station.

Three separate crews spent time in *Skylab* in 1973 and 1974. No further use was made of *Skylab* and it later fell out of orbit and burned up in the Earth's atmosphere.

At a reunion of *Skylab* astronauts in 1988, they were asked if they would go to Mars. The hands of the eight present all shot up at once.

△ An artist's impression of astronauts on Phebos, a moon of the planet Mars.

Mother in space

When Anna Fisher went up as mission specialist in the *Discovery* space shuttle in November 1984, she became the first mother to fly in space. Her husband William completed a family double when he also flew as a mission specialist in *Discovery*, in 1985.

27

Glossary

EVA
Extravehicular activity (EVA) means work outside a spacecraft; also called spacewalking.

Gravity
The pull exerted by one body on another. The Earth's gravitational pull keeps people and things on its surface. It requires massive rocket power for anything to escape.

Life-support system
Manned spacecraft are equipped with life-support systems that provide the occupants with all they need to live and breath — oxygen, food and water, and warmth. They also provide a hygienic method for getting rid of body wastes. Personal life-support systems are carried by astronauts on EVA.

Mission specialists
Astronauts qualified to operate a spacecraft's mission equipment, such as the robot arm on a shuttle, or to carry out tasks during EVA.

MMU
Manned Maneuvering Unit — a special backpack that enables an astronaut to operate in orbit untethered to the spacecraft.

Orbit
The path taken by one body around another. A spacecraft, for example, might orbit the Earth, being kept in orbit by the Earth's gravitational pull.

Payload specialists
Astronauts who carry out certain scientific experiments.

Weightlessness
People and objects on Earth or any other large body have weight — the force produced by the gravitational pull of that body. In orbit, away from Earth, astronauts feel as if they have no weight. This is called weightlessness.

Zero gravity
The condition that exists in space out of range of the pull of the Earth or any other large body.

Index

PRINTED IN BELGIUM BY
proost
INTERNATIONAL BOOK PRODUCTION